Narwhal

by Grace Hansen

Abdo Kids Jumbo is an Imprint of Abdo Kids
abdobooks.com

abdobooks.com

Published by Abdo Kids, a division of ABDO, P.O. Box 398166, Minneapolis, Minnesota 55439.
Copyright © 2020 by Abdo Consulting Group, Inc. International copyrights reserved in all countries. No part of this book may be reproduced in any form without written permission from the publisher.
Abdo Kids Jumbo™ is a trademark and logo of Abdo Kids.
Printed in the United States of America, North Mankato, Minnesota.

102019

012020

Photo Credits: Alamy, Animals Animals, iStock, Minden Pictures, National Geographic Image Collection, Seapics.com, Shutterstock

Production Contributors: Teddy Borth, Jennie Forsberg, Grace Hansen
Design Contributors: Dorothy Toth, Pakou Moua

Library of Congress Control Number: 2019941236

Publisher's Cataloging in Publication Data

Names: Hansen, Grace, author.

Title: Narwhal / by Grace Hansen

Description: Minneapolis, Minnesota : Abdo Kids, 2020 | Series: Arctic animals | Includes online resources and index.

Identifiers: ISBN 9781532188879 (lib. bdg.) | ISBN 9781532189364 (ebook) | ISBN 9781098200343 (Read-to-Me ebook)

Subjects: LCSH: Narwhal--Juvenile literature. | Marine mammals--Juvenile literature. | Whales--Juvenile literature. | Zoology--Arctic regions--Juvenile literature. | Marine animals--Adaptation--Arctic regions--Juvenile literature. | Arctic--Juvenile literature.

Classification: DDC 599.52--dc2

Table of Contents

The Arctic

The Arctic is the northernmost part of Earth. It is made up of land, the Arctic Ocean, and the **sea ice** that floats on it. The weather there is freezing cold. Any animal that lives in the Arctic is tough!

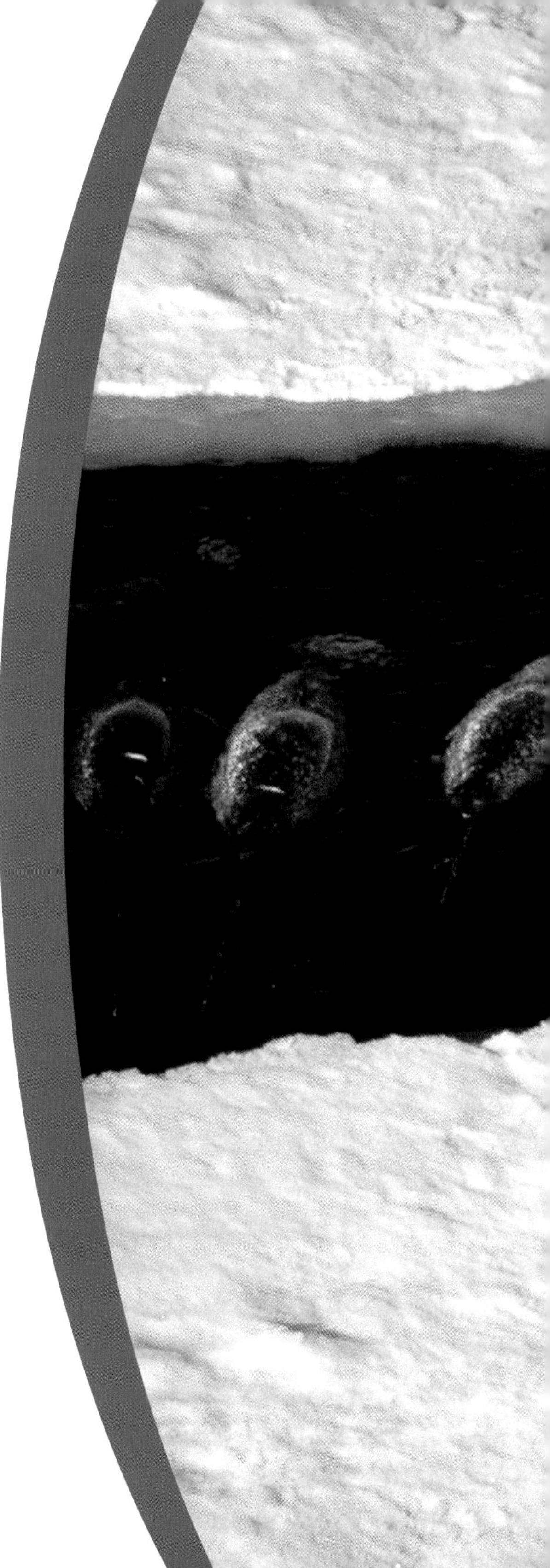

Narwhals

Narwhals live in the icy waters of the Arctic. Some people call them unicorns of the sea. But a narwhal is not a **myth**. It is a type of small, **toothed whale**.

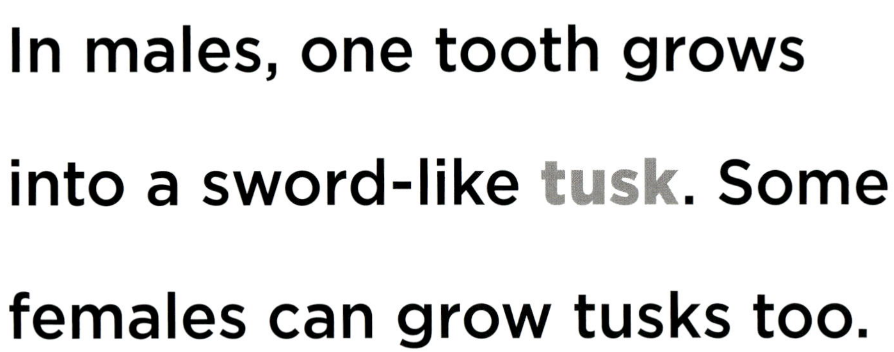

In males, one tooth grows into a sword-like **tusk**. Some females can grow tusks too.

The **tusk** is long and twisted. It can grow up to 10 feet (3 m) long!

Narwhals can grow to be 18 feet (5.5 m) long. They can weigh up to 4,200 pounds (1,905 kg).

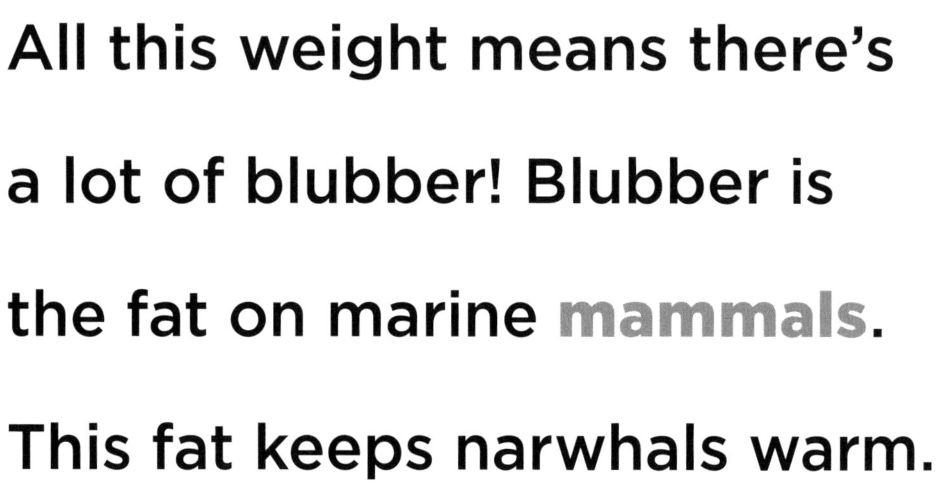

All this weight means there's a lot of blubber! Blubber is the fat on marine **mammals**. This fat keeps narwhals warm.

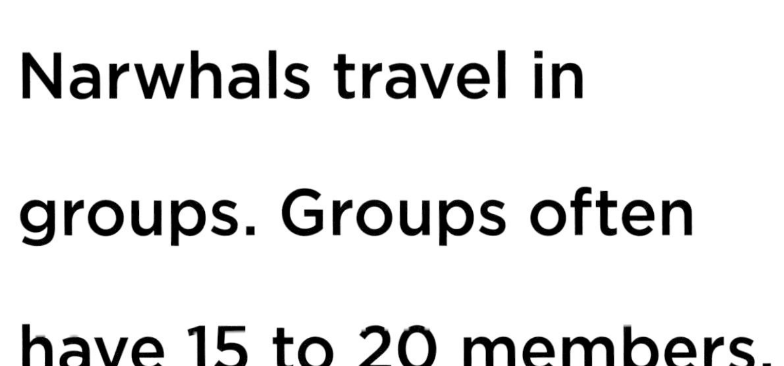

Narwhals travel in groups. Groups often have 15 to 20 members.

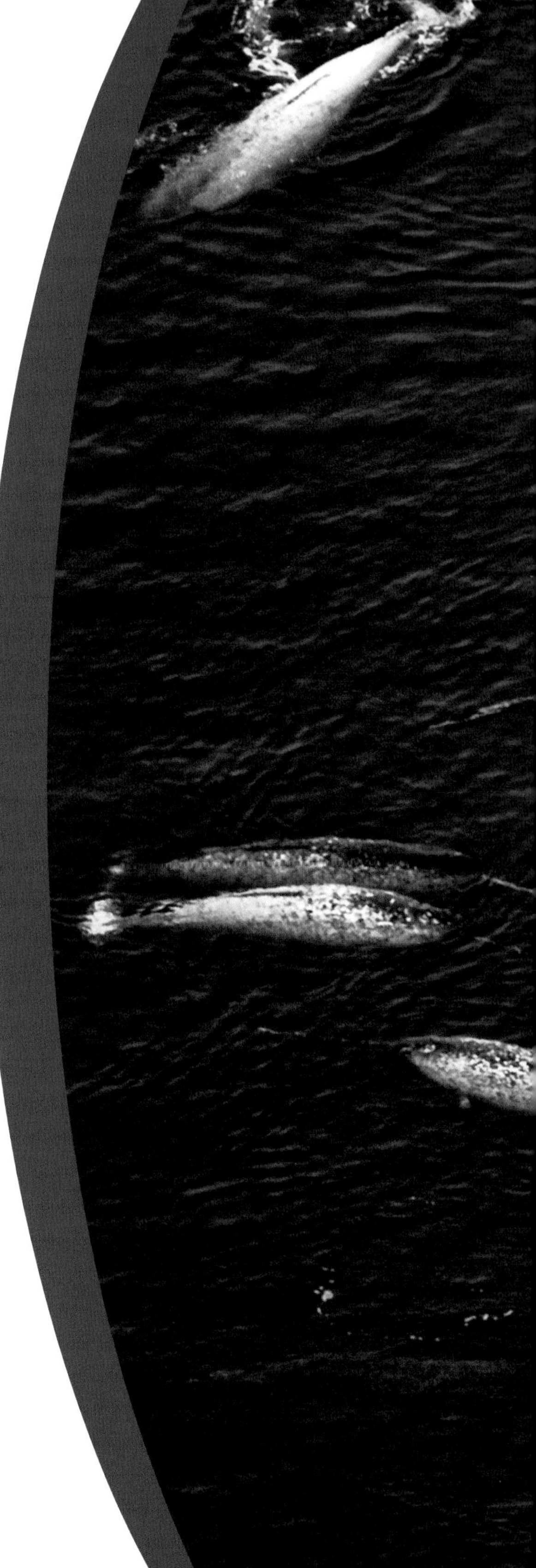

Narwhal groups travel to find food. They like to eat fish, like Greenland halibut and Arctic cod. Scientists think narwhals tap prey with their **tusks**. This stuns the prey so the narwhal can eat it.

Baby Narwhals

A female narwhal has one baby at a time. **Calves** are often born between June and August. They drink their mothers' milk for about 20 months. They grow and learn to hunt.

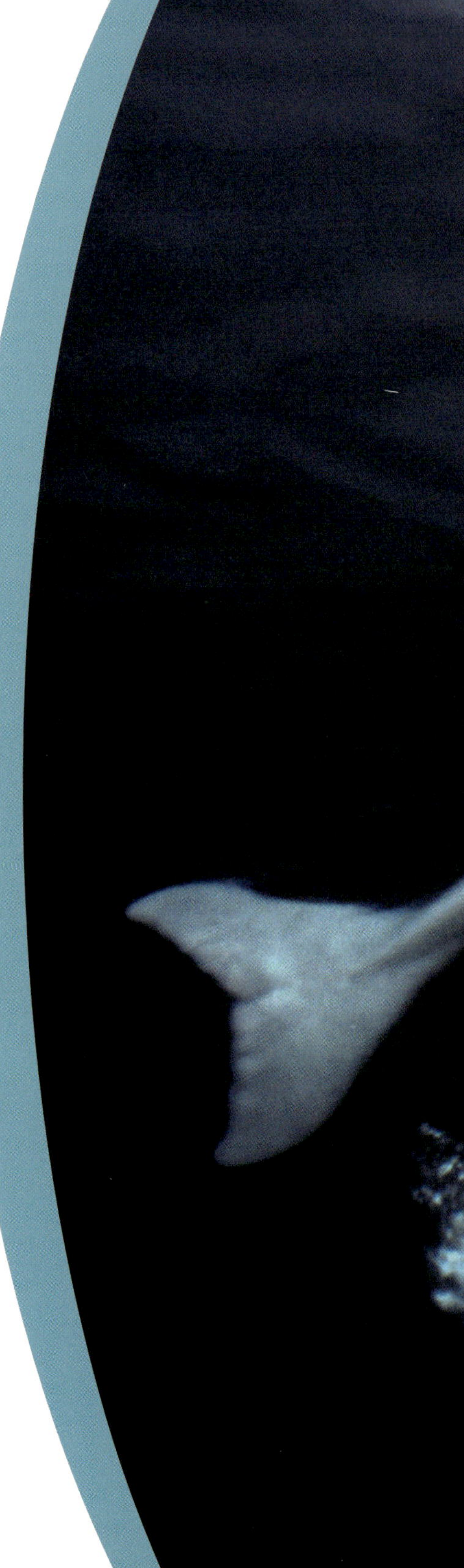

More Facts

- Even though a narwhal's **tusk** is a tooth, it does not grow like other teeth. The tusk grows through the narwhal's upper lip. This is why it looks like a horn.

- Besides fish, narwhals like to eat shrimp, squid, and other seafood.

- Narwhal groups are often small. But groups of hundreds have been seen in the wild.

Glossary

calf – the young of a whale and some other mammals.

mammal – a warm-blooded animal with fur or hair on its skin and a skeleton inside its body.

sea ice – frozen ocean water that is typically covered with snow.

toothed whale – a whale that has teeth instead of baleen.

tusk – a long, large pointed tooth that sticks out from the mouth of some animals.

Index

Visit abdokids.com

to access crafts, games, videos, and more!

Use Abdo Kids code

ANK8879

or scan this QR code!